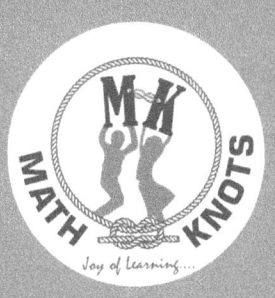

I0619965

This book belongs to

Name:_____

Cover Design by :
Gowri Vemuri

First Edition :
April , 2020

Author :
Gowri Vemuri

Edited by :
Raksha Pothapragada
Ritvik Pothapragada

Questions: mathknots.help@gmail.com

www.math-knots.com

INDEX

www.math-knots.com

Introduction

Basic rules of number system involves addition, subtraction , multiplication and division.
As the topic of "Numbers" involves some more useful concepts like LCM and GCD,
we shall study them in this chapter. For finding LCM and GCD, the divisibility rules are useful in one way or the other. Hence let us start the learning the divisibility rules.

Divisibility:

In general, if two natural numbers a and b are such that, when 'a' is divided by 'b', a remainder of zero is obtained, we say that 'a' is divisible by 'b'.

For example, 12 is divisible by 3 because 12 when divided by 3, the remainder is zero.

Also, we say that 12 is not divisible by 5, because 12 when divided by 5, it leaves a remainder 2.

Tests of Divisibility:

We now study the methods to test the divisibility of natural numbers with 2, 3, 4, 5, 6, 8, 9 and 11 without performing actual division.

Test of Divisibility by 2:

A natural number is divisible by 2, if its units digit is divisible by 2, i.e.,
the units place is either 0 or 2 or 4 or 6 or 8.

Examples : The numbers 4096, 23548 and 34052 are divisible by '2' as they end with 6, 8 and 2 respectively.

Test of Divisibility by 3:

A natural is divisible by 3 if the sum of its digits is divisible by 3.

Example: Consider the number 2143251. The sum of the digits of 2143251 $(2 + 1 + 4 + 3 + 2 + 5 + 1)$ is 18.

As 18 is divisible by 3, the number 2143251, is divisible by 3.

www.math-knots.com

Test of Divisibility by 4:

A natural number is divisible by 4, if the number formed by its last two digits is divisible by 4.

Examples : 4096, 53216, 548 and 4000 are all divisible by 4 as the numbers formed by taking the last two digits in each case is divisible by 4.

Test of Divisibility by 5:

A natural number is divisible by 5, if its units digit is either 0 or 5.

Examples : The numbers 4095 and 235060 are divisible by 5 as they have in their units place 5 and 0 respectively.

Test of Divisibility by 6:

A number is divisible by 6, if it is divisible by both 2 and 3.

Examples : Consider the number 753618

Since its units digit is 8, so it is divisible by 2. Also its sum of digits = 7 + 5 + 3 + 6 + 1 + 8 = 30, As 30 is divisible by 3, so 753618 is divisible by 3.
Hence 753618 is divisible by 6.

Test of Divisibility by 8:

A number is divisible by 8, if the number formed by its last three digits is divisible by 8.

Examples : 15840, 5432 and 7096 are all divisible by 8 as the numbers formed by last three digits in each case is divisible by 8.

Test of Divisibility by 9:

A natural number is divisible by 9, if the sum of its digits is divisible by 9.

Examples :

(i)   Consider the number 125847.
        Sum of digits = 1 + 2 + 5 + 8 + 4 + 7 = 27. As 27 is divisible by 9, the number 125847 is divisible by 9.

(ii)   Consider the number 145862.
        Sum of digits = 1 + 4 + 5 + 8 + 6 + 2 = 26. As 26 not divisible by 9, the number 145862 is not divisible by 9.

Factors and Exponents

Test of Divisibility by 11:

A number is divisible by 11, if the difference between the sum of the digits in odd places and sum of the digits in even places of the number is either 0 or a multiple of 11.

Examples :
(i)   Consider the number 9582540
       Now (sum of digits in odd places) - (sum of digits in even places)
       $= (9 + 8 + 5 + 0) - (5 + 2 + 4)$
       $= 11$, which is divisible by 11.
       Hence 958254 is divisible by 11.

(ii)   Consider the number 1453625
       Now, (sum of digits at odd places) - (sum of digits at even places)
       $= (1 + 5 + 6 + 5) - (4 + 3 + 2)$
       $= 17 - 9 = 8$, which is not divisible by 11

Some Additional Results :

If a natural number N is divisible by two natural numbers a and b, then N is divisible by the product of a and b, if and only if a and b are co primes.

Examples :
(i)   345 is divisible by 3 as well as by 5, as 3 and 5 are co-primes, 345 is divisible by 15.
(ii)   120 is divisible by 8 and 10, but it is not divisible by 80.

Factors and Multiples :

Lets learn the concepts of factors and multiples.

Definition :

If 'b' divides 'a' leaving a zero remainder, then 'b' is called a factor or divisor of 'a'
and 'a' is called the multiple of 'b'.
For example, $6 = 2 \ 3$
Here 2 and 3 are factors of 6 (or) 2 and 3 are divisors of 6.
And, 6 is a multiple of 3, 6 is a multiple of 2.

Examples :   (i)   The factors of $24 = \{1, 2, 3, 4, 6, 8, 12, 24\}$
                  (ii)   The factors of $256 = \{1, 2, 4, 8, 16, 32, 64, 128, 256\}$

Observations :

One is the factor of every natural number and it is the least of the factors of any natural number.
Every natural number is the factor of itself and it is the greatest.

Some Additional Results :

Unique Prime Factorisation Theorem : "Any natural number greater than 1 can be divided into a prime number or a composite number."

For example:

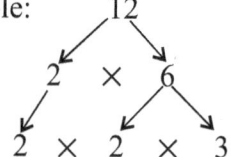

6 is a composite number

All factors are prime numbers.

If a is a composite number of the form $a = b^p c^q d^r$ ..... where b, c, d ... are district prime factors, then the number of factors of $a = (p+1)(q+1)(r+1)....$

If a is a composite number of the form $a = b^p c^q d^r$ .... where b, c, d .... are district prime factors, then the sum of all the factors of a $= \dfrac{(b^{p-1}-1)}{(b-1)} \times \dfrac{(c^{q-1}-1)}{(c-1)} \times \dfrac{(d^{r-1}-1)}{(d-1)}$ ......

Perfect Numbers :

A number for which sum of all its factors is twice the number itself is called a perfect number.

Observation :

Euler proved that if $2^k - 1$ is a prime number, then $2^{k-1}(2^k - 1)$ is a perfect number.
A perfect number can never be a prime number.

Examples :
(i)   Consider the composite number 6.
        Factors of 6 = {1, 2, 3, 6}
        Sum of factors = (1 + 2 + 3 + 6) = 12
        Clearly, the sum of the factors of 6 is twice the number itself.
        Hence 6 is a perfect number.

Example :
(ii)   Consider the composite number 48
        Factors of 48 = {1, 2, 3, 4, 6, 8, 12, 16, 24, 48}
        Sum of factors = 1 + 2 + 3 + 4 + 6 + 8 + 12 + 16 + 24 + 48
        = 124 2 48
        Clearly, 48 is not a perfect number.

Greatest Common Divisor [GCD] (or) Greatest Common Factor [GCF] (or) Highest Common Factor [HCF]

Definition :

"The greatest common factor of two or more natural numbers is the largest factor in the set of common factors of those numbers." In other words, the GCD (or) GCF of two or more numbers is the largest number that divides each of them exactly.

Example :   Find the GCF of 72 and 60.

Solution :   Let the set of factors of 72 be A.
A = {1, 2, 3, 4, 6, 8, 9, 12, 18, 24, 36, 72}
Let the set of factors of 60 be B.
B = {1, 2, 3, 4, 5, 6, 10, 12, 15, 20, 30, 60}
The set of common factors for 72 and 60 is A B = {1, 2, 3, 4, 6, 12}
The greatest element in this set is 12
The GCF (or) GCD for 72 and 60 is 2.

Observations :
If two numbers have no factors in common, then their GCF is unity.
i.e., GCF of prime numbers and co-prime numbers is unity.

Methods of finding GCF :

Factors Method :

When the numbers whose GCF has to be found are relatively small, this is the best suited method. Here we resolve the given numbers into their prime factors and find out the largest factor in the set of common factors to given numbers.

This method can be easily applied to any number of numbers.

Examples :   (i)   Find the GCF of 24 and 36.

Solution :

Resolving given numbers into product of prime factors, we have
36 = ②×②×③×3
24 = 2×②×②×③
The common factors to both the numbers are circled.
Now GCF = product common factors of given numbers = 2 × 2 × 3 = 12
GCF (24, 36) = 12

(ii)   Find the GCF of 12, 18 and 24.

Solution :

Resolving given numbers into product of prime factors;
12 = ② × 2 × ③
18 = ② × 3 × ③
24 = ② × 2 × 2 × ③

GCF   = product of common factors of 12, 18 and 24
        = 2 × 3 = 6
GCF   = 6

Division Method :

When the numbers whose GCF has to be found are very large, it is time consuming to write down sets of common factors to given numbers;

In this case, we use the method of Long Division. This method was proposed by Euclid and the following steps are involved in it.

Step 1 : Divide the larger number by the smaller number. If the remainder is zero, the divisor is the GCF, otherwise not.

Step 2 : Let the divisor in step 2 be the dividend now, and the remainder of step 1 become the divisor of step 2. Again, if the remainder is zero, the divisor is GCD. Otherwise, step 2 has to be repeated.

Example : Find the GCF of 64 and 56

Solution :
64 divided by 56, quotient is 1 and remainder is 8. Because the reminder is not zero; 56 is not the GCD.

Proceeding further, as mentioned in step 2, 56 is dividend and 8 is divisor. The quotient is 3 and remainder is zero.

Because the remainder is zero, the divisor 8 is the GCD.

GCF of three numbers using division method:

The GCF of 3 numbers is found out by finding GCF of any 2 numbers and GCF of the remaining number with the GCF obtained above.
i.e., GCF (a, b, c) = GCF [GCF (a, b), c].

www.math-knots.com

This process can be extended to any number of numbers.

Example : Find the GCF of 25, 45 and 75.

Solution : Let us first find the GCF of 25 and 45

```
25) 45 (1
    25
    20) 25 (1
        20
        5) 20 (4
           20
            0
```
0

GCF (25, 45) = 5
The GCF of (5, 75) is 5.
The GCF of 25, 45 and 75 is 5.

Observations :

The process of dividing the divisor with quotient is to be repeated until remainder as zero is obtained

If the zero remainder is obtained when the divisor is 1, then the GCD is '1'.

GCD is '1', means that two numbers are relatively prime or co-prime; because they do not have any factor in common other than 1. For eg; 12 and 13 are co-primes.

Some Additional Results :

The largest number which divides p, q and r to give remainders of s, t and u respectively will be the GCD of the three numbers $(p-s)$, $(q-t)$ and $(r-u)$.

The largest number which divides the numbers p, q and r and gives the same remainder in each case will be the GCD of the differences of two or the three numbers $(p-q, q-r, p-r)$.

Least Common Multiple [LCM] :

Definition :

"The least common multiple of two or more natural numbers is the least of their common multiples". In other words, the LCM of two or more numbers is the least number which can be divided exactly by each of the given numbers.

www.math-knots.com

Note : If the set of common multiples is denoted by C, then N and the number of elements in C is infinite and the least element in C is their LCM.

Example : Find L.C.M. of 24 and 36.

Solution : Resolving 24 and 36 into product of prime factors
$24 = ② × ② × 2 × ③$
$36 = ② × ② × ③ × 3$

The common prime factors of 24 and 36 are 2, 2 and 3. (which are circled)

The remaining prime factors of 24 is 2. (which is not circled).

The remaining prime factors of 36 is 3. (which is not circled).

LCM = Common factors of 24 the prime factors left in 24 the prime factors left in 36
$= 2 × 2 × 3 × 2 × 3$
$= 72$

Methods of finding LCM :

Factors Method :

Here the given numbers are decomposed into product of prime factors; from which, the least common multiple is found by multiplying the terms containing factors of numbers raised to their highest powers.

Example : Find the LCM of 32 and 24.

Solution : Resolving given numbers into product of common factors, we have
$32 = 2^5; 24 = 2^3 × 3$
LCM = product of terms containing highest powers of factors 2, 3
$= 2^5 \ 3 = 96$

LCM of three numbers using factors method:

The method above can be extended in a similar way to three numbers. This is illustrated below:

Example : Find the LCM of 12, 48 and 36

Solution : Resolving given numbers into product of common factors, we have
$12 = 2^2 × 3^1; 48 = 2^4 × 3^1; 36 = 2^2 × 3^2$. Then their LCM $= 2^2 × 3^2 = 16 × 9 = 144$

Synthetic Division Method of Finding LCM:

LCM of numbers can also be found using synthetic division method. This is illustrated below:

Example : Find the LCM of 144 and 156.

Solution : Using synthetic division, we have:

$$
\begin{array}{r|l}
2 & 144, 156 \\
\hline
2 & 72, 78 \\
\hline
3 & 36, 39 \\
\hline
& 12, 13
\end{array}
$$

LCM = 2 2 3 12 13
    = 1872

Examples : Find the LCM of 12, 18 and 24

Solution : Using synthetic division;

$$
\begin{array}{r|l}
2 & 12, 18, 24 \\
\hline
2 & 6, \; 9, \; 12 \\
\hline
2 & 3, \; 9, \; 6 \\
\hline
& 1, \; 3, \; 2
\end{array}
$$

LCM = 2 2 2 1 3 2 = 48

Some Additional Results :

Any natural N number which when divided by p, q or r leaving the same remainder s in each case will be of the form N = K (LCM of p, q, r) + s, where K = 0, 1, 2, 3 ....

Any natural number N which when divided by p, q and r leaves respective remainders of s, t and u where (p − s) = (q − t) = (r − u) = v (say), then it will be of the form
N = k (LCM of p, q and r) − v, where k = 1, 2, 3 ....

A natural N number which when divided by p and q leaves remainders r and s, will be of the form N = k (LCM of p and q) + n when n is the smallest integer solution for the equations n = pm_1 + r and n = qm_2 + s, where m_1 and m_2 are natural numbers.

15   www.math-knots.com

Relationship between LCM and GCF :

The LCM and GCF of two given numbers are related to the given numbers by the following relationship.
Product of the numbers = LCM × GCF

where, LCM denotes the LCM of the given numbers and GCF denotes the GCF of the given numbers.

Example :   Consider two numbers, 24 and 36.
These can be resolved into product of prime factors as below :
$24 = 2^3 \times 3$
$36 = 2^2 \times 3^2$
Now LCM $(24, 36) = 2^3 \times 3^2 = 72$
GCF $(24, 36) = 2^2 \times 3 = 12$

Now; Product of numbers $= 24 \times 36 = 2^5 \times 3^3 = 864$
Product of LCM and GCF $= 72 \times 12 = 2^5 \times 3^3 = 864$

Clearly, Product of the numbers = Product of the LCM and GCF.

Relatively Prime Numbers :

Definition :

If two numbers do not have any common factors other than 1, then they are called relatively prime numbers or co-prime numbers.

Concept : We know, every number has at least two factors, 1 and itself. If it has more than two factors, it is a composite number and if it does not have any factor except 1 and itself, it is a prime number.

But if two numbers (prime or composite) are such that they have only one common factor '1' are called relatively prime.

Example : Consider three numbers 8, 18 and 25
Now,
A, The set of factors of $8 = \{1, 2, 4, 8\}$
B, The set of factors of $18 = \{1, 2, 3, 6, 9, 18\}$
C, The set of factors of $25 = \{1, 5, 25\}$

Now, A B $= \{1, 2\}$; B C $= \{1\}$; C A $= \{1\}$

Clearly the common factors for both (18, 25) and (8, 25) is 1 only.

www.math-knots.com

They are generally written as $(18, 25) = 1$ and $(8, 25) = 1$

Note : Here; 18, 25 and 8 are not prime numbers (composite) but they are relatively prime numbers.

Observations :

The G.C.D. of two relatively prime numbers is 1 and their LCM is product of the numbers

Any two prime numbers are always relatively prime to each other.

Two relatively prime numbers need not be prime numbers.

LCM and GCF of Fractions :

The LCM and GCF of fractions can be determined by the following relations :

$$\text{LCM of fractions} = \frac{\text{LCM of numerators}}{\text{GCD of denominators}}$$

$$\text{GCF of fractions} = \frac{\text{GCD of numerators}}{\text{LCM of denominators}}$$

Examples : Find the GCD and LCM of $\frac{4}{5}$, $\frac{2}{5}$ and $\frac{3}{4}$

Solutions : $\text{LCM}\left(\frac{4}{5}, \frac{2}{5}, \frac{3}{4}\right) = \frac{\text{LCM}(4, 2, 3)}{\text{HCF}(5, 5, 4)} = \frac{4 \times 3}{5}$

$\text{GCD}\left(\frac{4}{5}, \frac{2}{5}, \frac{3}{4}\right) = \frac{\text{HCF}(4, 2, 3)}{\text{LCM}(5, 5, 4)} = \frac{1}{5 \times 4} = \frac{1}{20}$

www.math-knots.com

Introduction of Exponents

Whenever any integer, let us say x, is added n times, the result obtained will be equal to n times x i.e. nx. But in case, if the integer x is multiplied for n times, the result obtained will be equal to x^n (which is called as exponential form). The problems relating to these will be studied under 'Exponents'. We shall look at the rules/properties pertaining to these exponential numbers in this chapter.

Rational Exponents and Radicals

If 'a' is any real number and 'n' is a positive integer, then the product a × a × a × ---- n times is represented by the notation a^n. This notation is referred to as exponential form. In the above notation, a is called the base and n is called the power or exponent or index (plural of index is indices). a^n is read as 'nth power of a' or 'a to the power n'.

Example: $6 \times 6 \times 6 \times 6 \times 6 \times 6 \times 6$ can be written as 6^7. Here 6 is called as base and 7 is called as index (or exponent).

For a non-zero rational number 'a' with a negative integral exponent 'm' the following result can be observed.

$$a^m = a^{-n} = a^{-1} \times a^{-1} \times a^{-1} \times a^{-1} \times - - - - \times - - - \text{n times}$$

$$= \frac{1}{a} \times \frac{1}{a} \times \frac{1}{a} \times \frac{1}{a} - - - - \text{n times} = \left(\frac{1}{a}\right)^{-n} = \frac{1}{a^n}$$

Example: $6^3 = \left(\frac{1}{6}\right)^{-3}$

Rational Indices:

1. nth root of a :

A real number x is said to be the nth root of a if $x^n = a$; where a is any real number and n is a positive integer.

Parts of the exponent:

7^4

4 is the exponent

7 is the base

This is read as "Seven to the fourth power"

www.math-knots.com

The nth root of a can be represented as $a^{1/n}$ or $\sqrt[n]{a}$. Here $a^{1/n}$ is called exponential form and the form $\sqrt[n]{a}$ is called radical form. The sign $\sqrt[n]{}$ is called radical sign and $\sqrt[n]{a}$ is called radical. The number n is a positive integer is called the index of radical and a is called the radicand.

Example: We know that $32 = 2^5$. So we can say that 2 is 5th root of 32. It is written as $32^{1/5} = 2$ or $\sqrt[5]{32} = 2$

Similarly $\sqrt[3]{64} = 3$, $\sqrt[4]{625} = 5$, $\sqrt[6]{64} = 2$, etc.

Note:

(i) If n is negative as in case $64^{-\frac{1}{3}}$, we write the radical form as follows;

$$64^{-\frac{1}{3}} = \left(\frac{1}{64}\right)^{\frac{1}{3}} = \sqrt[3]{\frac{1}{64}} .$$

The radical form of $64^{-\frac{1}{3}}$ should not be taken as $\sqrt[3]{64}$ as in the radical form $\sqrt[n]{a}$ of n is a positive integer.

i.e., $a^{-1/n} = \sqrt[n]{\frac{1}{a}}$, where n is a positive integer.

(ii) $\sqrt[n]{a}$ is positive for a > 0 and n being a positive integer.
Example: $\sqrt[5]{32} = 2$, $\sqrt[6]{64} = 2$, $\sqrt[3]{27} = 3$, etc.

(iii) $\sqrt[n]{a}$ is negative for a < 0 and n being any odd positive integer.
Example: $\sqrt[3]{8} = -2$, $\sqrt[7]{128} = -2$, $\sqrt[5]{243} = -3$, etc.

(iv) $\sqrt[n]{a}$ does not exist in set of real numbers, for a < 0 and n being even positive integer.
Example: $\sqrt[2]{16}$, $\sqrt[4]{256}$, $\sqrt[6]{64}$ etc doesn't exist.

Each positive number has two square roots, one positive and the other negative.

Example: $\sqrt{36} = 6$ or -6 (since $6^2 = (-6)^2 = 36$).

If a is a positive rational number and n = p/q is a positive rational exponent, then we can define $a^{p/q}$ in two ways.

(1) $a^{\frac{p}{q}}$ is the qth root of a^p, i.e. $a^{\frac{p}{q}} = \left(a^p\right)^{\frac{1}{q}}$

(2) $a^{\frac{p}{q}}$ is the pth power of qth root of a, i.e. $a^{\frac{p}{q}} = \left(a^{\frac{1}{q}}\right)^p$.

Laws of indices:

For all real numbers a and b and all rational numbers m and n, we have

(i) $a^m \times a^n = a^{m+n}$

   Examples: (1) $2^3 \times 2^6 = 2^{3+6} = 2^9$

   (2) $\left(\dfrac{5}{6}\right)^4 \times \left(\dfrac{5}{6}\right)^5 = \left(\dfrac{5}{6}\right)^{4+5} = \left(\dfrac{5}{6}\right)^9$

   (3) $5^{2/3} \times 5^{4/3} = 5^{(2/3+4/3)} = 5^{6/3} = 5^2$

   (4) $2^3 \times 2^4 \times 2^5 \times 2^8 = 2^{(3+4+5+8)} = 2^{20}$.

   (5) $\left(\sqrt{7}\right)^3 \times \left(\sqrt{7}\right)^{\frac{5}{2}} = \left(\sqrt{7}\right)^{3+\frac{5}{2}} = \left(\sqrt{7}\right)^{\frac{11}{2}}$

(ii) $a^m \div a^n = a^{m-n}, a \neq 0$

   Examples: (a) $7^8 \div 7^3 = 7^{8-3} = 7^5$

   (b) $\left(\dfrac{7}{3}\right)^9 \div \left(\dfrac{7}{3}\right)^5 = \left(\dfrac{7}{3}\right)^{9-5} = \left(\dfrac{7}{3}\right)^4$

   (c) $9^{\frac{2}{3}} \div 9^{\frac{1}{6}} = 9^{\left(\frac{2}{3}-\frac{1}{6}\right)} = 9^{\left(\frac{4-1}{6}\right)} = 9^{\frac{3}{6}} = 9^{\frac{1}{2}}$

   (d) $\left(\dfrac{5}{7}\right)^{\frac{8}{9}} \div \left(\dfrac{5}{7}\right)^{\frac{1}{3}} = \left(\dfrac{5}{7}\right)^{\left(\frac{8}{9}-\frac{1}{3}\right)} = \left(\dfrac{5}{7}\right)^{\frac{8-3}{9}} = \left(\dfrac{5}{7}\right)^{\frac{5}{9}}$

www.math-knots.com

Note: $a^n \div a^n = 1$
or $a^{n-n} = a^0 = 1$
$\therefore a^0 = 1, a \neq 0$

(iii) $(a^m)^n = a^{m \times n}$

Examples: (a) $(5^2)^3 = 5^{2 \times 3} = 5^6$

(b) $\left[\left(\dfrac{2}{3}\right)^4\right]^5 = \left(\dfrac{2}{3}\right)^{4 \times 5} = \left(\dfrac{2}{3}\right)^{20}$

(c) $\left[\left(\dfrac{5}{7}\right)^{\frac{2}{3}}\right]^{\frac{9}{8}} = \left(\dfrac{5}{7}\right)^{\left(\frac{2}{3} \times \frac{9}{8}\right)} = \left(\dfrac{5}{7}\right)^{\frac{3}{4}}$

(iv) $\left(\dfrac{a}{b}\right)^n = \dfrac{a^n}{b^n}$

Example: $\left(\dfrac{4}{5}\right)^7 = \dfrac{4^7}{5^7}$

Note: Conversely we can write $\left(\dfrac{a^n}{b^n}\right) = \left(\dfrac{a}{b}\right)^n$

Example: $\dfrac{8}{27} = \dfrac{2^3}{3^3} = \left(\dfrac{2}{3}\right)^3$

(v) $(ab)^n = a^n \times b^n$

Examples: (a) $20)^5 = (4 \times 5)^5 = 4^5 \times 5^5$
(b) $(42)^7 = (2 \times 3 \times 7)^7 = 2^7 \times 3^7 \times 7^7$

Note: Conversely we can write $a^n \times b^n = (ab)^n$

www.math-knots.com

Examples: (a) $4^8 \times 5^8 = (4 \times 5)^8 = 20^8$

(b) $\left(\dfrac{2}{3}\right)^5 \times \left(\dfrac{9}{8}\right)^5 = \left(\dfrac{2}{3} \times \dfrac{9}{8}\right)^5 = \left(\dfrac{3}{4}\right)^5$

(vi) $a^{-n} = \dfrac{1}{a^n}$, $a \neq 0$

Example: $2^{-4} = \dfrac{1}{2^4}$, $5^{-1} = \dfrac{1}{5}$

Note: $a^{-1} = \dfrac{1}{a^1} = \dfrac{1}{a}$

(vii) $\left(\dfrac{a}{b}\right)^n = \left(\dfrac{b}{a}\right)^n$

Examples: (a) $\left(\dfrac{5}{9}\right)^3 = \left(\dfrac{9}{5}\right)^{-3}$

(b) $\left(\dfrac{1}{5}\right)^{-1} = \left(\dfrac{5}{1}\right)^1 = 5$

Note: $\left(\dfrac{1}{a}\right)^{-1} = \left(\dfrac{a}{1}\right)^1 = a$

(viii) If $a^m = a^n$, then m = n, where $a \neq 0$, $a \neq 1$

Examples: (a) If $5^p = 5^3 \Rightarrow p = 3$

(b) If $4^p = 256$

$4^p = 4^4 \Rightarrow p = 4$

(ix) For positive numbers a and b, if $a^n = b^n$, $n \neq 0$, then a = b (when n is odd)

Examples: (a) If $5^7 = p^7$, then clearly p = 5.

(b) If $(5)^{2n-1} = (3 \times p)^{2n-1}$, then clearly 5 = 3p or p = 5/3

www.math-knots.com

(x)   If $p^m \times q^n \times r^s = p^a\, q^b\, r^c$, then m = a, n = b, s = c, where p, q, r are different primes.

Examples: (a) If $40500 = 2^a \times 5^b \times 3^c$, then find $a^a \times b^b \times c^c$

2	40,500
2	20,250
5	10,125
5	2,025
5	405
3	81
3	27
3	9
	3

$\therefore 40500 = 2^2 \times 5^3 \times 3^4 = 2^a \times 5^b \times 3^c$

$\therefore a = 2, b = 3, c = 4$, [Using the above law].

$\therefore a^a \times b^b \times c^c = 2^2 \times 3^3 \times 4^4 = 27{,}648$

Example 8 :   (a) $20)^5 = (4 \times 5)^5 = 4^5 \times 5^5$

             (b) $(42)^7 = (2 \times 3 \times 7)^7 = 2^7 \times 3^7 \times 7^7$

Note: Conversely we can write $a^n \times b^n = (ab)^n$

Example 9 :   (a) $\left((5)^3\right)^2 = 5^{3\times2} = (5)^6 = 5\times5\times5\times5\times5\times5 = 15625$

(b) $(2)^3 \times (2)^5 = (2)^{3+5} = (2)^8 = 2\times2\times2\times2\times2\times2\times2\times2 = 256$

(c) $(7)^0 = 1$ | Any base value rise to the power zero is always equal to 1 |

(d) $(3)^{-4} = \dfrac{1}{(3)^4} = \dfrac{1}{3\times3\times3\times3} = \dfrac{1}{81}$

(d) $\dfrac{(8)^7}{(8)^5} = (8)^{7-5} = (8)^2 = 64$

(e) $\dfrac{(9)^4}{(9)^7} = (9)^{4-7} = (9)^{-3} = \dfrac{1}{(9)^3} = \dfrac{1}{9\times9\times9} = \dfrac{1}{729}$

(f) $(2)^4 = 2\times2\times2\times2 = 16$ (g) $(-2)^4 = -2\times-2\times-2\times-2 = 16$

(h) $-(2)^4 = -(2\times2\times2\times2) = -16$ (i) $-(2)^3 = -(2\times2\times2) = -8$

(j) $(-2)^3 = -2\times-2\times-2 = -8$ (k) $-(-2)^3 = -(-2\times-2\times-2) = -(-8) = 8$

Tip 1 : When the exponent is an even number the simplified value is always positive, when the base has a positive or negative value.

Tip 2 : When the exponent is an odd number the simplified value is always positive, when the base has a positive value.

Tip 3 : When the exponent is an odd number the simplified value is always negative, when the base has a negative value.

Factors and
Exponents

Positive Factors

List all the positive factors for each of the numbers given below :

(1) 207

(2) 344

(3) 272

(4) 337

(5) 370

(6) 204

(7) 273

(8) 353

(9) 331

(10) 332

List all the positive factors for each of the numbers given below :

(11) 329

(12) 341

(13) 238

(14) 395

(15) 206

(16) 391

(17) 276

(18) 388

(19) 237

(20) 289

List all the positive factors for each of the numbers given below :

(21) 366

(22) 284

(23) 327

(24) 274

(25) 296

(26) 356

(27) 306

(28) 376

(29) 202

(30) 291

www.math-knots.com

List all the positive factors for each of the numbers given below :

(31) 279

(32) 200

(33) 218

(34) 230

(35) 322

(36) 294

(37) 368

(38) 354

(39) 304

(40) 385

List all the positive factors for each of the numbers given below :

(41) 347

(42) 297

(43) 361

(44) 225

(45) 358

(46) 367

(47) 247

(48) 213

(49) 382

(50) 220

www.math-knots.com

List all the positive factors for each of the numbers given below :

(51) 226

(52) 209

(53) 246

(54) 305

(55) 215

(56) 235

(57) 228

(58) 363

(59) 259

(60) 271

www.math-knots.com

List all the positive factors for each of the numbers given below :

(61) 293

(62) 340

(63) 320

(64) 308

(65) 262

(66) 318

(67) 219

(68) 212

(69) 310

(70) 278

List all the positive factors for each of the numbers given below :

(71) 268

(72) 328

(73) 393

(74) 227

(75) 392

(76) 372

(77) 349

(78) 334

(79) 346

(80) 239

List all the positive factors for each of the numbers given below :

(81) 397

(82) 307

(83) 394

(84) 233

(85) 298

(86) 309

(87) 303

(88) 333

(89) 398

(90) 383

Prime Factorization

Write the prime factorization without using exponents for each of the numbers given below :

(91) 267

(92) 273

(93) 375

(94) 321

(95) 300

(96) 255

(97) 205

(98) 296

(99) 392

(100) 376

www.math-knots.com

Write the prime factorization without using exponents for each of the numbers given below :

(101) 301

(102) 222

(103) 278

(104) 276

(105) 216

(106) 262

(107) 357

(108) 314

(109) 305

(110) 235

Write the prime factorization without using exponents for each of the numbers given below :

(111) 285

(112) 256

(113) 224

(114) 220

(115) 302

(116) 400

(117) 237

(118) 394

(119) 340

(120) 289

38

www.math-knots.com

Write the prime factorization without using exponents for each of the numbers given below :

(121) 287

(122) 214

(123) 399

(124) 298

(125) 372

(126) 308

(127) 230

(128) 252

(129) 270

(130) 343

Write the prime factorization without using exponents for each of the numbers given below :

(131) 387   (132) 346

(133) 244   (134) 221

(135) 247   (136) 272

(137) 351   (138) 368

(139) 290   (140) 322

www.math-knots.com

Write the prime factorization without using exponents for each of the numbers given below :

(141) 236

(142) 384

(143) 354

(144) 282

(145) 378

(146) 393

(147) 260

(148) 295

(149) 238

(150) 365

Write the prime factorization without using exponents for each of the numbers given below :

(151) 219

(152) 388

(153) 243

(154) 286

(155) 371

(156) 345

(157) 309

(158) 318

(159) 364

(160) 380

www.math-knots.com

Write the prime factorization without using exponents for each of the numbers given below :

(161) 350

(162) 259

(163) 312

(164) 356

(165) 316

(166) 202

(167) 323

(168) 279

(169) 206

(170) 329

Write the prime factorization without using exponents for each of the numbers given below :

(171) 385

(172) 240

(173) 330

(174) 258

(175) 315

(176) 386

(177) 398

(178) 336

(179) 304

(180) 234

44

www.math-knots.com

Prime Power Factorization

Write the prime factors as exponents where applicable for each of the numbers given below :

(181) 212

(182) 203

(183) 336

(184) 218

(185) 228

(186) 396

(187) 344

(188) 291

(189) 287

(190) 289

www.math-knots.com

Write the prime factors as exponents where applicable for each of the numbers given below :

(191) 286

(192) 256

(193) 298

(194) 258

(195) 321

(196) 314

(197) 306

(198) 372

(199) 201

(200) 214

www.math-knots.com

Write the prime factors as exponents where applicable for each of the numbers given below :

(201) 270

(202) 200

(203) 388

(204) 393

(205) 255

(206) 242

(207) 268

(208) 318

(209) 231

(210) 325

www.math-knots.com

Write the prime factors as exponents where applicable for each of the numbers given below :

(211) 297

(212) 395

(213) 390

(214) 282

(215) 303

(216) 222

(217) 345

(218) 328

(219) 248

(220) 341

48

www.math-knots.com

Write the prime factors as exponents where applicable for each of the numbers given below :

(221) 362

(222) 250

(223) 261

(224) 267

(225) 240

(226) 265

(227) 230

(228) 278

(229) 378

(230) 309

Write the prime factors as exponents where applicable for each of the numbers given below :

(231) 274

232) 216

(233) 210

234) 369

(235) 338

(236) 361

(237) 326

(238) 206

(239) 276

(240) 387

www.math-knots.com

Write the prime factors as exponents where applicable for each of the numbers given below :

(241) 329

(242) 364

(243) 334

(244) 399

(245) 357

(246) 232

(247) 385

(248) 346

(249) 320

(250) 285

Write the prime factors as exponents where applicable for each of the numbers given below :

(251) 363

(252) 295

(253) 375

(254) 304

(255) 252

(256) 302

(257) 322

(258) 351

(259) 315

(260) 370

www.math-knots.com

Write the prime factors as exponents where applicable for each of the numbers given below :

(261) 294

(262) 335

(263) 332

(264) 330

(265) 356

(266) 253

(267) 386

(268) 365

(269) 398

(270) 234

Greatest Common Factors

Find the Greatest Common Factor (GCF) for each of the numbers given below :

(271) 16, 40, 24

(272) 24, 32, 40

(273) 36, 8, 24

(274) 18, 12, 21

(275) 33, 15, 9

(276) 22, 24, 20

(277) 33, 36, 3

(278) 39, 9, 6

(279) 36, 12, 32

(280) 40, 28, 20

www.math-knots.com

Find the Greatest Common Factor (GCF) for each of the numbers given below :

(281) 22, 30, 16

(282) 38, 23, 26

(283) 22, 12, 38

(284) 39, 35, 27

(285) 10, 30, 25

(286) 18, 24, 27

(287) 24, 18, 6

(288) 39, 21, 9

(289) 6, 9, 33

(290) 36, 30, 12

Find the Greatest Common Factor (GCF) for each of the numbers given below :

(291) 24, 6, 9

(292) 27, 21, 39

(293) 15 10 35

(294) 6 10 24

(295) 32, 16, 28

(296) 24, 27, 21

(297) 39, 15, 9

(298) 16, 20, 24

(299) 21, 6, 12

(300) 35, 40, 20

www.math-knots.com

Find the Greatest Common Factor (GCF) for each of the numbers given below :

(301) 24, 32, 16

(302) 30, 33, 18

(303) 18, 24, 12

(304) 36, 39, 6

(305) 12, 25, 18

(306) 8, 16, 12

(307) 18, 12, 30

(308) 40, 32, 16

(309) 36, 18, 12

(310) 18, 39, 12

Find the Greatest Common Factor (GCF) for each of the numbers given below :

(311) 6, 21, 27

(312) 33, 27, 15

(313) 20, 8, 40

(314) 28, 21, 14

(315) 30, 9, 36

(316) 21, 35, 28

(317) 20, 36, 24

(318) 20, 22, 36

(319) 30, 24, 18

(320) 30, 9, 24

Find the Greatest Common Factor (GCF) for each of the numbers given below :

(321) 30, 26, 6

(322) 10, 25, 20

(323) 11, 3, 39

(324) 12, 24, 8

(325) 8, 18, 34

(326) 30, 12, 39

(327) 8, 32, 12

(328) 15, 30, 35

(329) 40, 20, 25

(330) 30, 12, 24

www.math-knots.com

Find the Greatest Common Factor (GCF) for each of the numbers given below :

(331) 30, 36, 18

(332) 25, 30, 15

(333) 6, 16, 30

(334) 14, 35, 21

(335) 27, 36, 18

(336) 12, 16, 28

(337) 39, 6, 27

(338) 9, 36, 27

(339) 38, 14, 32

(340) 40, 10, 25

www.math-knots.com

Find the Greatest Common Factor (GCF) for each of the numbers given below :

(341) 36, 32, 40

(342) 10, 20, 15

(343) 36, 40, 24

(344) 8, 12, 20

(345) 28, 8, 40

(346) 24, 30, 36

(347) 10, 30, 4

(348) 10, 18, 8

(349) 36, 24, 12

(350) 35, 28, 14

www.math-knots.com

Find the Greatest Common Factor (GCF) for each of the numbers given below :

(351) 26, 40, 22

(352) 30, 40, 20

(353) 30, 39, 15

(354) 8, 36, 32

(355) 25, 15, 20

(356) 17, 70, 84

(357) 71, 24, 76

(358) 42, 36, 78

(359) 72, 36, 90

(360) 66, 22, 55

Find the Greatest Common Factor (GCF) for each of the numbers given below :

(361) 52, 26, 65

(362) 75, 70, 89

(363) 60, 45, 30

(364) 55, 40, 25

(365) 91, 52, 65

(366) 100, 50, 75

(367) 26, 98, 12

(368) 96, 36, 60

(369) 85, 55, 75

(370) 31, 8, 96

Find the Greatest Common Factor (GCF) for each of the numbers given below :

(371) 72, 24, 12

(372) 50, 25, 35

(373) 54, 90, 81

(374) 72, 36, 96

(375) 100, 40, 80

(376) 60, 70, 50

(377) 80, 40, 12

(378) 46, 69, 92

(379) 85, 51, 68

(380) 30, 66, 78

Find the Greatest Common Factor (GCF) for each of the numbers given below :

(381) 57, 84, 81

(382) 60, 70, 30

(383) 42, 90, 96

(384) 78, 60, 81

(385) 60, 40, 100

(386) 88, 64, 16

(387) 95, 38, 76

(388) 88, 99, 77

(389) 54, 90, 36

(390) 85, 90, 20

www.math-knots.com

Find the Greatest Common Factor (GCF) for each of the numbers given below :

(391) 64, 96, 80

(392) 55, 77, 66

(393) 10, 95, 50

(394) 36, 18, 96

(395) 9, 36, 42

(396) 45, 90, 54

(397) 47, 41, 34

(398) 80, 70, 60

(399) 16, 8, 48

(400) 72, 56, 64

www.math-knots.com

Find the Greatest Common Factor (GCF) for each of the numbers given below :

(401) 70, 91, 28

(402) 78, 81, 30

(403) 91, 52, 78

(404) 6, 51, 21

(405) 88, 66, 44

(406) 60, 40, 80

(407) 90, 54, 72

(408) 57, 76, 38

(409) 60, 96, 72

(410) 45, 39, 33

Find the Greatest Common Factor (GCF) for each of the numbers given below :

(411) 24, 32, 88

(412) 85, 34, 68

(413) 65, 91, 39

(414) 88, 40, 24

(415) 80, 32, 48

(416) 63, 42, 84

(417) 36, 60, 84

(418) 30, 80, 50

(419) 36, 84, 72

(420) 77, 56, 42

Find the Greatest Common Factor (GCF) for each of the numbers given below :

(421) 36, 90, 84

(422) 36, 72, 48

(423) 48, 96, 36

(424) 32, 40, 48

(425) 63, 21, 14

(426) 35, 94, 77

(427) 90, 81, 99

(428) 39, 65, 26

(429) 24, 8, 72

(430) 64, 96, 32

Find the Greatest Common Factor (GCF) for each of the numbers given below :

(431) 12, 30, 54

(432) 48, 96, 72

(433) 63, 77, 35

(434) 98, 14, 63

(435) 27, 72, 36

(436) 60, 27, 45

(437) 60, 92, 32

(438) 18, 63, 90

(439) 90, 30, 45

(440) 45, 85, 70

www.math-knots.com

Find the Greatest Common Factor (GCF) for each of the numbers given below :

(441) 54, 126, 72

(442) 180, 90, 135

(443) 78, 143, 104

(444) 56, 80, 200

(445) 96, 64, 192

(446) 198, 66, 165

(447) 40, 20, 28

(448) 92, 184, 69

(449) 84, 42, 63

(450) 130, 182, 78

Find the Greatest Common Factor (GCF) for each of the numbers given below :

(451) 150, 102, 180

(452) 195, 75, 150

(453) 152, 57, 171

(454) 140, 175, 70

(455) 66, 110, 154

(456) 146, 198, 152

(457) 180, 54, 126

(458) 52, 182, 130

(459) 180, 200, 40

(460) 102, 170, 136

Find the Greatest Common Factor (GCF) for each of the numbers given below :

(461) 116, 58, 174

(462) 78, 39, 130

(463) 160, 155, 190

(464) 120, 80, 200

(465) 162, 135, 81

(466) 104, 78, 130

(467) 84, 189, 168

(468) 136, 68, 51

(469) 186, 62, 93

(470) 184, 64, 120

Find the Greatest Common Factor (GCF) for each of the numbers given below :

(471) 72, 36, 60

(472) 117, 195, 65

(473) 192, 32, 176

(474) 120, 24, 36

(475) 86, 172, 129

(476) 16, 144, 40

(477) 136, 192, 112

(478) 192, 128, 160

(479) 111, 74, 185

(480) 120, 72, 168

www.math-knots.com

Find the Greatest Common Factor (GCF) for each of the numbers given below :

(481) 136, 34, 119

(482) 153, 27, 30

(483) 195, 156, 117

(484) 92, 138, 184

(485) 162, 54, 90

(486) 24, 98, 162

(487) 105, 30, 135

(488) 80, 100, 150

(489) 120, 60, 200

(490) 8, 96, 176

Find the Greatest Common Factor (GCF) for each of the numbers given below :

(491) 147, 196, 98

(492) 111, 185, 148

(493) 200, 137, 157

(494) 189, 135, 81

(495) 135, 45, 165

(496) 52, 130, 104

(497) 75, 165, 180

(498) 44, 143, 121

(499) 39, 52, 182

(500) 80, 90, 100

76   www.math-knots.com

Find the Greatest Common Factor (GCF) for each of the numbers given below :

(501) 38, 190, 57

(502) 96, 72, 144

(503) 97, 139, 68

(504) 100, 200, 150

(505) 134, 120, 95

(506) 144, 90, 36

(507) 160, 128, 176

(508) 132, 72, 60

(509) 4, 89, 16

(510) 176, 132, 88

www.math-knots.com

Find the Greatest Common Factor (GCF) for each of the numbers given below :

(511) 192, 96, 144

(512) 68, 102, 170

(513) 38, 133, 152

(514) 115, 46, 161

(515) 160, 120, 200

(516) 90, 18, 72

(517) 66, 154, 88

(518) 136, 68, 170

(519) 56, 154, 70

(520) 120, 150, 60

www.math-knots.com

Find the Greatest Common Factor (GCF) for each of the numbers given below :

(521) 76, 114, 190

(522) 84, 112, 49

(523) 180, 90, 120

(524) 123, 82, 164

(525) 54, 81, 189

Least Common Multiple

Find the Least Common Multiple (LCM) for each of the numbers given below :

(526) 36, 16

(527) 18, 33

(528) 32, 28

(529) 16, 26

Find the Least Common Multiple (LCM) for each of the numbers given below :

(530) 16, 34

(531) 12, 27

(532) 16, 24

(533) 16, 28

(534) 26, 30

(535) 18, 36

(536) 15, 21

(537) 15, 35

(538) 30, 20

(539) 24, 32

www.math-knots.com

Find the Least Common Multiple (LCM) for each of the numbers given below :

(540) 36, 40

(541) 27, 18

(542) 40, 16

(543) 16, 38

(544) 30, 24

(545) 38, 20

(546) 10, 18

(547) 12, 24

(548) 38, 36

(549) 15, 6

Find the Least Common Multiple (LCM) for each of the numbers given below :

(550) 9, 39

(551) 38, 32

(552) 4, 30

(553) 36, 15

(554) 20, 15

(555) 20, 32

(556) 26, 12

(557) 18, 12

(558) 40, 25

(559) 9, 10

www.math-knots.com

Find the Least Common Multiple (LCM) for each of the numbers given below :

(560) 20, 28

(561) 28, 35

(562) 26, 18

(563) 21, 28

(564) 30, 12

(565) 36, 27

(566) 20, 24

(567) 8, 36

(568) 36, 4

(569) 35, 25

www.math-knots.com

Find the Least Common Multiple (LCM) for each of the numbers given below :

(570) 15, 12

(571) 15, 33

(572) 15, 40

(573) 8, 12

(574) 24, 40

(575) 35, 21

(576) 4, 18

(577) 24, 39

(578) 24, 36

(579) 33, 24

www.math-knots.com

Find the Least Common Multiple (LCM) for each of the numbers given below :

(580) 39, 35

(581) 40, 20

(582) 18, 24

(583) 30, 38

(584) 14, 36

(585) 35, 30

(586) 25, 15

(587) 25, 30

(588) 20, 35

(589) 38, 40

www.math-knots.com

Find the Least Common Multiple (LCM) for each of the numbers given below :

(590) 36, 28

(591) 33, 6

(592) 14, 35

(593) 32, 12

(594) 21, 33

(595) 34, 18

(596) 27, 24

(597) 24, 22

(598) 8, 26

(599) 21, 30

www.math-knots.com

Find the Least Common Multiple (LCM) for each of the numbers given below :

(600) 30, 36

(601) 40, 32

(602) 27, 30

(603) 14, 21

(604) 6, 39

(605) 21, 27

(606) 18, 15

(607) 24, 14

(608) 40, 30

(609) 14, 28

Find the Least Common Multiple (LCM) for each of the numbers given below :

(610) 18, 30

(611) 60, 90

(612) 20, 70

(613) 80, 32

(614) 96, 64

(615) 85, 51

(616) 44, 88

(617) 50, 30

(618) 50, 75

(619) 25, 65

www.math-knots.com

Find the Least Common Multiple (LCM) for each of the numbers given below :

(620) 40, 58

(621) 56, 48

(622) 78, 65

(623) 72, 48

(624) 91, 26

(625) 39, 78

(626) 99, 51

(627) 48, 80

(628) 39, 52

(629) 88, 98

www.math-knots.com

Find the Least Common Multiple (LCM) for each of the numbers given below :

(630) 72, 24

(631) 14, 35

(632) 42, 12

(633) 84, 63

(634) 57, 76

(635) 90, 36

(636) 65, 55

(637) 28, 42

(638) 60, 24

(639) 33, 88

www.math-knots.com

Find the Least Common Multiple (LCM) for each of the numbers given below :

(640) 35, 60

(641) 88, 24

(642) 38, 95

(643) 99, 63

(644) 42, 98

(645) 100, 80

(646) 20, 50

(647) 90, 99

(648) 33, 68

(649) 66, 44

www.math-knots.com

Find the Least Common Multiple (LCM) for each of the numbers given below :

(650) 100, 40

(651) 64, 48

(652) 70, 56

(653) 54, 72

(654) 48, 60

(655) 38, 57

(656) 60, 93

(657) 91, 65

(658) 76, 95

(659) 60, 100

Find the Least Common Multiple (LCM) for each of the numbers given below :

(660) 63, 35

(661) 84, 70

(662) 36, 96

(663) 84, 60

(664) 36, 40

(665) 84, 32

(666) 90, 72

(667) 100, 75

(668) 45, 81

(669) 56, 24

Find the Least Common Multiple (LCM) for each of the numbers given below :

(670) 56, 82

(671) 90, 75

(672) 80, 60

(673) 90, 54

(674) 48, 84

(675) 63, 42

(676) 90, 20

(677) 91, 78

(678) 24, 16

(679) 20, 64

www.math-knots.com

Find the Least Common Multiple (LCM) for each of the numbers given below :

(680) 58, 22

(681) 80, 50

(682) 96, 80

(683) 32, 56

(684) 34, 98

(685) 76, 32

(686) 30, 96

(687) 72, 81

(688) 96, 72

(689) 26, 8

www.math-knots.com

Find the Least Common Multiple (LCM) for each of the numbers given below :

(690) 56, 84

(691) 34, 51

(692) 88, 99

(693) 92, 69

(694) 88, 66

(695) 60, 16

(696) 84, 168, 126

(697) 86, 172, 129

(698) 48, 96, 120

(699) 173, 105, 35

www.math-knots.com

Find the Least Common Multiple (LCM) for each of the numbers given below :

(700) 152, 114, 190

(701) 22, 99, 77

(702) 100, 160, 140

(703) 80, 48, 160

(704) 120, 48, 144

(705) 81, 135, 189

(706) 168, 72, 120

(707) 168, 192, 48

(708) 184, 36, 16

(709) 50, 175, 150

Find the Least Common Multiple (LCM) for each of the numbers given below :

(710) 121, 110, 187

(711) 168, 56, 112

(712) 117, 156, 195

(713) 167, 59, 190

(714) 132, 198, 165

(715) 42, 98, 70

(716) 169, 52, 26

(717) 160, 96, 80

(718) 125, 75, 93

(719) 132, 88, 176

Find the Least Common Multiple (LCM) for each of the numbers given below :

(720) 195, 120, 150

(721) 125, 193, 147

(722) 90, 45, 105

(723) 182, 147, 21

(724) 132, 54, 6

(725) 107, 84, 16

(726) 69, 27, 129

(727) 98, 63, 91

(728) 100, 200, 150

(729) 185, 148, 74

www.math-knots.com

Find the Least Common Multiple (LCM) for each of the numbers given below :

(730) 90, 135, 180

(731) 36, 81, 99

(732) 142, 39, 72

(733) 200, 40, 60

(734) 125, 200, 150

(735) 180, 60, 120

(736) 132, 33, 121

(737) 168, 196, 56

(738) 135, 120, 60

(739) 186, 155, 93

www.math-knots.com

Find the Least Common Multiple (LCM) for each of the numbers given below :

(740) 140, 70, 105

(741) 125, 160, 175

(742) 164, 12, 48

(743) 160, 50, 200

(744) 156, 78, 117

(745) 145, 185, 40

(746) 160, 128, 96

(747) 68, 187, 85

(748) 132, 86, 178

(749) 24, 84, 120

www.math-knots.com

Find the Least Common Multiple (LCM) for each of the numbers given below :

(750) 84, 91, 168

(751) 55, 165, 134

(752) 54, 189, 162

(753) 15, 95, 150

(754) 176, 143, 187

(755) 15, 200, 145

(756) 170, 68, 102

(757) 128, 96, 64

(758) 141, 94, 188

(759) 78, 195, 117

Find the Least Common Multiple (LCM) for each of the numbers given below :

(760) 160, 176, 64

(761) 104, 65, 195

(762) 90, 162, 27

(763) 182, 104, 78

(764) 180, 90, 150

(765) 58, 145, 87

(766) 69, 27, 189

(767) 90, 120, 150

(768) 152, 76, 114

(769) 155, 93, 124

www.math-knots.com

Find the Least Common Multiple (LCM) for each of the numbers given below :

(770) 88, 66, 176

(771) 200, 160, 80

(772) 96, 144, 192

(773) 81, 67, 46

(774) 133, 171, 38

(775) 144, 123, 189

(776) 54, 72, 36

(777) 196, 147, 98

(778) 28, 70, 98

(779) 132, 154, 44

Find the Least Common Multiple (LCM) for each of the numbers given below :
(780) 87, 28, 170

<u>Simplification of exponents</u>

Simplify the below exponents. Write them as positive exponents.

(781) $11^6 \cdot 11^7$

(782) $3 \cdot 3^{10}$

(783) $9 \cdot 9^3$

(784) $9^3 \cdot 9^0 \cdot 9^8$

(785) $15^5 \cdot 15^8$

(786) $4^5 \cdot 4^2$

(787) $18^4 \cdot 18^9$

(788) $18^7 \cdot 18^4$

www.math-knots.com

Simplify the below exponents. Write them as positive exponents.

(789) $13^9 \cdot 13^2$

(790) $13 \cdot 13^3$

(791) $18^{10} \cdot 18^8$

(792) $3^5 \cdot 3^{10}$

(793) $2^6 \cdot 2^4$

(794) $4^6 \cdot 4^7$

(795) $19^4 \cdot 19^3$

(796) $19^3 \cdot 19^5$

(797) $13^8 \cdot 13^7$

(798) $10^2 \cdot 10^9$

www.math-knots.com

Simplify the below exponents. Write them as positive exponents.

(799) $3^0 \cdot 3^6$

(800) $19 \cdot 19^5$

(801) $14^0 \cdot 14^2$

(802) $11^3 \cdot 11^8$

(803) $17^3 \cdot 17^6$

(804) $7 \cdot 7^5 \cdot 7^{10}$

(805) $19 \cdot 19^3$

(806) $7^4 \cdot 7^6$

(807) $17^8 \cdot 17^4 \cdot 17^0$

(808) $4 \cdot 4^0$

Simplify the below exponents. Write them as positive exponents.

(809) $15^6 \cdot 15^4$

(810) $19^5 \cdot 19^9$

(811) $11^4 \cdot 11^9 \cdot 11^0$

(812) $15^2 \cdot 15^8$

(813) $6^5 \cdot 6^5$

(814) $25 \cdot 5^7$

(815) $7^8 \cdot 7^{10}$

(816) $7^6 \cdot 7^{10}$

(817) $3^9 \cdot 3^8$

(818) $10^7 \cdot 10^3$

Simplify the below exponents. Write them as positive exponents.

(819) $12^0 \cdot 12^2$

(820) $20 \cdot 20^4$

(821) $19 \cdot 19^8$

(822) $18^4 \cdot 18^0 \cdot 18^{10}$

(823) $13^5 \cdot 13^6$

(824) $14 \cdot 14^3$

(825) $19^3 \cdot 19^9$

(826) $17^5 \cdot 17^7$

(827) $2^9 \cdot 2^3$

(828) $8^3 \cdot 8^{10}$

www.math-knots.com

Simplify the below exponents. Write them as positive exponents.

(829) $19^8 \cdot 19^3$

(830) $11 \cdot 11^9$

(831) $20^8 \cdot 20^8 \cdot 20^4$

(832) $14^0 \cdot 14^7$

(833) $19^6 \cdot 19^8 \cdot 19^0$

(834) $2^0 \cdot 2^7$

(835) $8^8 \cdot 8^0$

(836) $2^2 \cdot 2^9 \cdot 2^4$

(837) $8^2 \cdot 8^0$

(838) $6 \cdot 6^2$

Simplify the below exponents. Write them as positive exponents.

(839) $6^{10} \cdot 6^4$

(840) $8^7 \cdot 8^0$

(841) $12^4 \cdot 12^4$

(842) $2^8 \cdot 2^6$

(843) $12^{10} \cdot 12^3$

(844) $11^4 \cdot 11^2$

(845) $2^0 \cdot 2^6$

(846) $4^2 \cdot 4^7 \cdot 4^3$

(847) $10^8 \cdot 10^5$

(848) $2^3 \cdot 2^6$

Simplify the below exponents. Write them as positive exponents.

(849) $11^0 \cdot 11^6$

(850) $12^5 \cdot 12^4$

(851) $15 \cdot 15^8 \cdot 15^7$

(852) $5 \cdot 5^6$

(853) $13^9 \cdot 13^4$

(854) $17^6 \cdot 17^8$

(855) $5^7 \cdot 5^3$

(856) $14^{10} \cdot 14^5$

(857) $4^9 \cdot 4^8$

(858) $19^4 \cdot 19^8$

Simplify the below exponents. Write them as positive exponents.

(859) $2 \cdot 2^4$

(860) $6 \cdot 6^0 \cdot 6^9$

(861) $4^7 \cdot 4^3$

(862) $11^7 \cdot 11^7$

(863) $2^8 \cdot 2^{10}$

(864) $9 \cdot 9^0$

(865) $4^4 \cdot 4^2$

(866) $\left(3^4\right)^3$

(867) $\left(8^4\right)^0$

(868) $\left(8^2\right)^3$

www.math-knots.com

Simplify the below exponents. Write them as positive exponents.

(869) $\left(5^3\right)^1$

(870) $\left(8^2\right)^2$

(871) $\left(2^3\right)^3$

(872) $\left(8^3\right)^1$

(873) $\left(4^3\right)^2$

(874) $\left(5^4\right)^2$

(875) $\left(7^3\right)^2$

(876) $\left(6^3\right)^4$

(877) $\left(6^3\right)^3$

(878) $\left(3^4\right)^2$

www.math-knots.com

Simplify the below exponents. Write them as positive exponents.

(879) $\left(4^3\right)^4$

(880) $\left(6^3\right)^2$

(881) $\left(7^4\right)^2$

(882) $\left(7^3\right)^1$

(883) $\left(2^4\right)^2$

(884) $\left(3^3\right)^3$

(885) $\left(5^2\right)^3$

(886) $\left(3^2\right)^2$

(887) $\left(8^4\right)^1$

(888) $\left(5^2\right)^2$

Simplify the below exponents. Write them as positive exponents.

(889) $\left(2^3\right)^1$

(890) $\left(2^4\right)^3$

(891) $\left(2^3\right)^2$

(892) $\left(7^3\right)^3$

(893) $\left(7^2\right)^2$

(894) $\left(5^4\right)^3$

(895) $\left(6^4\right)^2$

(896) $\left(6^2\right)^1$

(897) $\left(3^3\right)^4$

(898) $\left(3^2\right)^4$

Simplify the below exponents. Write them as positive exponents.

(899) $\left(8^3\right)^3$

(900) $\left(3^3\right)^1$

(901) $\left(5^3\right)^4$

(902) $\left(4^2\right)^0$

(903) $\left(2^2\right)^3$

(904) $\left(3^3\right)^2$

(905) $\left(8^2\right)^4$

(906) $\left(7^2\right)^1$

(907) $\left(5^3\right)^0$

(908) $\left(4^0\right)^2$

Simplify the below exponents. Write them as positive exponents.

(909) $\left(6^3\right)^1$

(910) $\left(4^4\right)^1$

(911) $\left(3^2\right)^3$

(912) $\left(5^2\right)^4$

(913) $\left(4^2\right)^4$

914) $\left(6^0\right)^1$

(915) $\left(6^4\right)^1$

(916) $\left(5^3\right)^2$

(917) $\left(6^2\right)^3$

(918) $\left(2^4\right)^1$

Simplify the below exponents. Write them as positive exponents.

(919) $\left(5^4\right)^1$

(920) $\left(4^2\right)^2$

(921) $\left(3^2\right)^1$

(922) $\left(7^4\right)^1$

(923) $\left(3^4\right)^4$

(924) $\left(6^2\right)^2$

(925) $\left(7^2\right)^4$

(926) $\left(7^2\right)^3$

(927) $\left(6^2\right)^4$

(928) $\left(8^4\right)^3$

www.math-knots.com

Simplify the below exponents. Write them as positive exponents.

(929) $\left(8^3\right)^2$

(930) $\left(2^2\right)^1$

(931) $\left(3^4\right)^1$

(932) $\left(4^4\right)^3$

(933) $\left(4^3\right)^1$

(934) $\left(2^4\right)^4$

(935) $\left(5^3\right)^3$

(936) $\dfrac{2^0}{2^{10}}$

(937) $\dfrac{12^6}{12^9}$

(938) $\dfrac{3^5}{3^9}$

Simplify the below exponents. Write them as positive exponents.

(939) $\dfrac{10^3}{10}$

(940) $\dfrac{4^3}{4^4}$

(941) $\dfrac{13^7}{13^6}$

(942) $\dfrac{7^{10}}{7^6}$

(943) $\dfrac{14}{14^0}$

(944) $\dfrac{10^4}{10}$

(945) $\dfrac{17^7}{17^{10}}$

(946) $\dfrac{15^0}{15}$

(947) $\dfrac{9^{10}}{9^0}$

(948) $\dfrac{7^3}{7^9}$

www.math-knots.com

Simplify the below exponents. Write them as positive exponents.

(949) $\dfrac{2^{10}}{2^8}$

(950) $\dfrac{8^8}{8^8}$

(951) $\dfrac{14^5}{14^5}$

(952) $\dfrac{2^7}{2^8}$

(953) $\dfrac{10^{10}}{10^7}$

(954) $\dfrac{19^4}{19^2}$

(955) $\dfrac{18^6}{18^3}$

(956) $\dfrac{10^8}{10^9}$

(957) $\dfrac{16^6}{16^7}$

(958) $\dfrac{10^6}{10^0}$

Simplify the below exponents. Write them as positive exponents.

(959) $\dfrac{15^3}{15^{10}}$

(960) $\dfrac{12^3}{12^2}$

(961) $\dfrac{20^3}{20^4}$

(962) $\dfrac{11^6}{11^{10}}$

(963) $\dfrac{17^4}{17^6}$

(964) $\dfrac{19^6}{19^8}$

(965) $\dfrac{4^6}{4^6}$

(966) $\dfrac{14}{14^2}$

(967) $\dfrac{20^{10}}{20}$

(968) $\dfrac{20^6}{20^5}$

Simplify the below exponents. Write them as positive exponents.

(969) $\dfrac{8}{8^{10}}$

(970) $\dfrac{12}{12^8}$

(971) $\dfrac{10^2}{10^5}$

(972) $\dfrac{14^8}{14}$

(973) $\dfrac{9^{10}}{9}$

(974) $\dfrac{19}{19^2}$

(975) $\dfrac{10^9}{10^4}$

(976) $\dfrac{4^4}{4^2}$

(977) $\dfrac{2^0}{2^5}$

(978) $\dfrac{17^4}{17}$

Simplify the below exponents. Write them as positive exponents.

(979) $\dfrac{2^0}{2}$

(980) $\dfrac{11^8}{11^3}$

(981) $\dfrac{20^8}{20^9}$

(982) $\dfrac{14^0}{14^4}$

(983) $\dfrac{10^6}{10^3}$

(984) $\dfrac{2^2}{2^7}$

(985) $\dfrac{8^9}{8^7}$

(986) $\dfrac{7^{10}}{7^4}$

(987) $\dfrac{8^{10}}{8^8}$

(988) $\dfrac{10^5}{10}$

Simplify the below exponents. Write them as positive exponents.

(989) $\dfrac{14^0}{14^8}$

(990) $\dfrac{16^5}{16^6}$

(991) $\dfrac{3^6}{3^0}$

(992) $\dfrac{5^8}{5^6}$

(993) $\dfrac{6^0}{6^4}$

(994) $\dfrac{19^4}{19^{10}}$

(995) $\dfrac{10}{10^2}$

(996) $\dfrac{15^5}{15^{10}}$

(997) $\dfrac{12^0}{12^6}$

(998) $\dfrac{12^9}{12^{10}}$

Simplify the below exponents. Write them as positive exponents.

(999) $\dfrac{19^6}{19^7}$

(1000) $\dfrac{4}{4^3}$

(1001) $\dfrac{4^{10}}{4^9}$

(1002) $\dfrac{2}{2^2}$

(1003) $\dfrac{20}{20^5}$

(1004) $\dfrac{4^9}{4^0}$

(1005) $\dfrac{12^5}{12^2}$

(1006) $\dfrac{15^{10}}{15^{10}}$

(1007) $\dfrac{20^8}{20^{10}}$

(1008) $\dfrac{5}{5^4}$

www.math-knots.com

Simplify the below exponents. Write them as positive exponents.

(1009) $\dfrac{19^3}{19^4}$

(1010) $\dfrac{16^2}{16}$

(1011) $\dfrac{13^5}{13^8}$

(1012) $\dfrac{5}{5^0}$

(1013) $\dfrac{13^5}{13^3}$

(1014) $\dfrac{11^8}{11^{10}}$

(1015) $\dfrac{8^9}{8}$

(1016) $\dfrac{5^0}{5^{10}}$

(1017) $\dfrac{10^0}{10^5}$

(1018) $\dfrac{13^{10}}{13^3}$

www.math-knots.com

Simplify the below exponents. Write them as positive exponents.

(1019) $\dfrac{4^0}{4^3}$

(1020) $\dfrac{3^0}{3^7}$

www.math-knots.com

Answer Keys

www.math-knots.com

Answer Key

(1)   1, 3, 9, 23, 69, 207

(2)   1, 2, 4, 8, 43, 86, 172, 344

(3)   1, 2, 4, 8, 16, 17, 34, 68, 136, 272

(4)   1, 337

(5)   1, 2, 5, 10, 37, 74, 185, 370

(6)   1,2,3,4,6,12,17,34,51,68,102,204

(7)   1, 3, 7, 13, 21, 39, 91, 273

(8)   1, 353

(9)   1, 331

(10)   1, 2, 4, 83, 166, 332

(11)   1, 7, 47, 329

(12)   1, 11, 31, 341

(13)   1, 2, 7, 14, 17, 34, 119, 238

(14)   1, 5, 79, 395

(15)   1, 2, 103, 206

(16)   1, 17, 23, 391

(17)   1,2,3,4,6,12,23,46,69,92,138,276

(18)   1, 2, 4, 97, 194, 388

(19)   1, 3, 79, 237

(20)   1, 17, 289

(21)   1, 2, 3, 6, 61, 122, 183, 366

(22)   1, 2, 4, 71, 142, 284

(23)   1, 3, 109, 327

(24)   1, 2, 137, 274

(25)   1, 2, 4, 8, 37, 74, 148, 296

(26)   1, 2, 4, 89, 178, 356

(27)   1,2,3,6,9,17,18,34,51,102,153,306

(28)   1, 2, 4, 8, 47, 94, 188, 376

(29)   1, 2, 101, 202

(30)   1, 3, 97, 291

(31)   1, 3, 9, 31, 93, 279

(32)   1,2,4,5,8,10,20,25,40,50,100,200

(33)   1, 2, 109, 218

(34)   1, 2, 5, 10, 23, 46, 115, 230

(35)   1, 2, 7, 14, 23, 46, 161, 322

(36) 1, 2, 3, 6, 7, 14, 21, 42, 49, 98, 147, 294

(37) 1,2,4,8,16,23,46,92,184,368

(38) 1, 2, 3, 6, 59, 118, 177, 354

(39) 1,2,4,8,16,19,38,76,152,304

(40) 1, 5, 7, 11, 35, 55, 77, 385

(41) 1, 347

(42) 1, 3, 9, 11, 27, 33, 99, 297

(43) 1, 19, 361

(44) 1, 3, 5, 9, 15, 25, 45, 75, 225

(45) 1, 2, 179, 358

(46) 1, 367

(47) 1, 13, 19, 247

(48) 1, 3, 71, 213

(49) 1, 2, 191, 382

(50) 1, 2, 4, 5, 10, 11, 20, 22, 44, 55, 110, 220

(51) 1, 2, 113, 226

(52) 1, 11, 19, 209

(53) 1, 2, 3, 6, 41, 82, 123, 246

(54) 1, 5, 61, 305

(55) 1, 5, 43, 215

(56) 1, 5, 47, 235

(57) 1,2,3,4,6,12,19,38,57,76,114,228

(58) 1, 3, 11, 33, 121, 363

(59) 1, 7, 37, 259

(60) 1, 271

(61) 1, 293

(62) 1, 2, 4, 5, 10, 17, 20, 34, 68, 85, 170, 340

(63) 1,2,4,5,8,10,16,20,32,40,64,80,160,320

(65) 1, 2, 131, 262

(64) 1, 2, 4, 7, 11, 14, 22, 28, 44, 77, 154, 308

(67) 1, 3, 73, 219

(66) 1, 2, 3, 6, 53, 106, 159, 318

(68) 1, 2, 4, 53, 106, 212

(69)   1, 2, 5, 10, 31, 62, 155, 310

(70)   1, 2, 139, 278

(71)   1, 2, 4, 67, 134, 268

(72)   1, 2, 4, 8, 41, 82, 164, 328

(73)   1, 3, 131, 393

(74)   1, 227

(75)   1,2,4,7,8,14,28,49,56,98,196,392

(76)   1,2,3,4,6,12,31,62,93,124,186,372

(77)   1, 349

(78)   1, 2, 167, 334

(79)   1, 2, 173, 346

(80)   1, 239

(81)   1, 397

(82)   1, 307

(83)   1, 2, 197, 394

(84)   1, 233

(85)   1, 2, 149, 298

(86)   1, 3, 103, 309

(87)   1, 3, 101, 303

(88)   1, 3, 9, 37, 111, 333

(89)   1, 2, 199, 398

(90)   1, 383

(91) $3 \cdot 89$

(92) $3 \cdot 7 \cdot 13$

(93) $3 \cdot 5 \cdot 5 \cdot 5$

(94) $3 \cdot 107$

(95) $2 \cdot 2 \cdot 3 \cdot 5 \cdot 5$

(96) $3 \cdot 5 \cdot 17$

(97) $5 \cdot 41$

(98) $2 \cdot 2 \cdot 2 \cdot 37$

(99) $2 \cdot 2 \cdot 2 \cdot 7 \cdot 7$

(100) $2 \cdot 2 \cdot 2 \cdot 47$

(101) $7 \cdot 43$

(102) $2 \cdot 3 \cdot 37$

www.math-knots.com

(103)　2 · 139

(104)　2 · 2 · 3 · 23

(105)　2 · 2 · 2 · 3 · 3 · 3

(106)　2 · 131

(107)　3　7　17

(108)　2　157

(109)　5　61

(110)　5　47

(111)　3 · 5 · 19

(112)　2 · 2 · 2 · 2 · 2 · 2 · 2 · 2

(113)　2 · 2 · 2 · 2 · 2 · 7

(114)　2 · 2 · 5 · 11

(115)　2 · 151

(116)　2 · 2 · 2 · 2 · 5 · 5

(117)　3 · 79

(118)　2 · 197

(119)　2 · 2 · 5 · 17

(120)　17 · 17

(121)　7 · 41

(122)　2 · 107

(123)　3 · 7 · 19

(124)　2 · 149

(125)　2 · 2 · 3 · 31

(126)　2 · 2 · 7 · 11

(127)　2 · 5 · 23

(128)　2 · 2 · 3 · 3 · 7

(129)　2 · 3 · 3 · 3 · 5

(130)　7 · 7 · 7

(131)　3 · 3 · 43

(132)　2 · 173

(133)　2 · 2 · 61

(134)　13 · 17

(135)　13 · 19

(136)　2 · 2 · 2 · 2 · 17

(137) $3 \cdot 3 \cdot 3 \cdot 13$

(138) $2 \cdot 2 \cdot 2 \cdot 2 \cdot 23$

(139) $2 \cdot 5 \cdot 29$

(140) $2 \cdot 7 \cdot 23$

(141) $2 \cdot 2 \cdot 59$

(142) $2 \cdot 2 \cdot 2 \cdot 2 \cdot 2 \cdot 2 \cdot 2 \cdot 3$

(152) $2 \cdot 2 \cdot 97$

(143) $2 \cdot 3 \cdot 59$

(144) $2 \cdot 3 \cdot 47$

(145) $2 \cdot 3 \cdot 3 \cdot 3 \cdot 7$

(146) $3 \cdot 131$

(147) $2 \cdot 2 \cdot 5 \cdot 13$

(148) $5 \cdot 59$

(149) $2 \cdot 7 \cdot 17$

(150) $5 \cdot 73$

(151) $3 \cdot 73$

(153) $3 \cdot 3 \cdot 3 \cdot 3 \cdot 3$

(154) $2 \cdot 11 \cdot 13$

(155) $7 \cdot 53$

(156) $3 \cdot 5 \cdot 23$

(157) $3 \cdot 103$

(158) $2 \cdot 3 \cdot 53$

(159) $2 \cdot 2 \cdot 7 \cdot 13$

(160) $2 \cdot 2 \cdot 5 \cdot 19$

(161) $2 \cdot 5 \cdot 5 \cdot 7$

(162) $7 \cdot 37$

(163) $2 \cdot 2 \cdot 2 \cdot 3 \cdot 13$

(164) $2 \cdot 2 \cdot 89$

(165) $2 \cdot 2 \cdot 79$

(166) $2 \cdot 101$

(167) $17 \cdot 19$

(168) $3 \cdot 3 \cdot 31$

(169) $2 \cdot 103$

(170) $7 \cdot 47$

(171) $5 \cdot 7 \cdot 11$

(172) $2 \cdot 2 \cdot 2 \cdot 2 \cdot 3 \cdot 5$

(173) $2 \cdot 3 \cdot 5 \cdot 11$

(174) $2 \cdot 3 \cdot 43$

(175) $3 \cdot 3 \cdot 5 \cdot 7$

(176) $2 \cdot 193$

(177) $2 \cdot 199$

(178) $2 \cdot 2 \cdot 2 \cdot 2 \cdot 3 \cdot 7$

(179) $2 \cdot 2 \cdot 2 \cdot 2 \cdot 19$

(180) $2 \cdot 3 \cdot 3 \cdot 13$

(181) $2^2 \cdot 53$

(182) $7 \cdot 29$

(183) $2^4 \cdot 3 \cdot 7$

(184) $2 \cdot 109$

(185) $2^2 \cdot 3 \cdot 19$

(186) $2^2 \cdot 3^2 \cdot 11$

(187) $2^3 \cdot 43$

(188) $3 \cdot 97$

(189) $7 \cdot 41$

(190) 17^2

(191) $2 \cdot 11 \cdot 13$

(192) 2^8

(193) $2 \cdot 149$

(194) $2 \cdot 3 \cdot 43$

(195) $3 \cdot 107$

(196) $2 \cdot 157$

(197) $2 \cdot 3^2 \cdot 17$

(198) $2^2 \cdot 3 \cdot 31$

(199) $3 \cdot 67$

(200) $2 \cdot 107$

(201) $2 \cdot 3^3 \cdot 5$

(202) $2^3 \cdot 5^2$

(203) $2^2 \cdot 97$

(204) $3 \cdot 131$

(205) $3 \cdot 5 \cdot 17$

(206) $2 \cdot 11^2$

(207) $2^2 \cdot 67$

(208) $2 \cdot 3 \cdot 53$

(209) $3 \cdot 7 \cdot 11$

(210) $5^2 \cdot 13$

(211) $3^3 \cdot 11$

(212) $5 \cdot 79$

(213) $2 \cdot 3 \cdot 5 \cdot 13$

(214) $2 \cdot 3 \cdot 47$

(215) $3 \cdot 101$

(216) $2 \cdot 3 \cdot 37$

(217) $3 \cdot 5 \cdot 23$

(218) $2^3 \cdot 41$

(219) $2^3 \cdot 31$

(220) $11 \cdot 31$

(221) $2 \cdot 181$

(222) $2 \cdot 5^3$

(223) $3^2 \cdot 29$

(224) $3 \cdot 89$

(225) $2^4 \cdot 3 \cdot 5$

(226) $5 \cdot 53$

(227) $2 \cdot 5 \cdot 23$

(228) $2 \cdot 139$

(229) $2 \cdot 3^3 \cdot 7$

(230) $3 \cdot 103$

(231) $2 \cdot 137$

(232) $2^3 \cdot 3^3$

(233) $2 \cdot 3 \cdot 5 \cdot 7$

(234) $3^2 \cdot 41$

(235) $2 \cdot 13^2$

(236) 19^2

(237) $2 \cdot 163$

(238) $2 \cdot 103$

(239) $2^2 \cdot 3 \cdot 23$

(240) $3^2 \cdot 43$

(241) $7 \cdot 47$

(242) $2^2 \cdot 7 \cdot 13$

(243) $2 \cdot 167$

(244) $3 \cdot 7 \cdot 19$

(245) $3 \cdot 7 \cdot 17$

(246) $2^3 \cdot 29$

(247) $5 \cdot 7 \cdot 11$

(248) $2 \cdot 173$

(249) $2^6 \cdot 5$

(250) $3 \cdot 5 \cdot 19$

(251) $3 \cdot 11^2$

(252) $5 \cdot 59$

(253) $3 \cdot 5^3$

(254) $2^4 \cdot 19$

(255) $2^2 \cdot 3^2 \cdot 7$

(256) $2 \cdot 151$

(257) $2 \cdot 7 \cdot 23$

(258) $3^3 \cdot 13$

(259) $3^2 \cdot 5 \cdot 7$

(260) $2 \cdot 5 \cdot 37$

(261) $2 \cdot 3 \cdot 7^2$

(262) $5 \cdot 67$

(263) $2^2 \cdot 83$

(264) $2 \cdot 3 \cdot 5 \cdot 11$

(265) $2^2 \cdot 89$

(266) $11 \cdot 23$

(267) $2 \cdot 193$

(268) $5 \cdot 73$

(269) $2 \cdot 199$

(270) $2 \cdot 3^2 \cdot 13$

(271) 8

(272) 8

273) 4

274) 3

275) 3

276) 2

www.math-knots.com

(277)   3   (278)   3   (279)   4   (280)   4

(281)   2   (282)   1   (283)   2   (284)   1

(285)   5   (286)   3   (287)   6   (288)   3

(289)   3   (290)   6   (291)   3   (292)   3

(293)   5   (294)   2   (295)   4   (296)   3

(297)   3   (298)   4   (299)   3   (300)   5

(301)   8   (302)   3   (303)   6   (304)   3

(305)   1   (306)   4   (307)   6   308)   8

(309)   6   (310)   3   (311)   3   (312)   3

(313)   4   (314)   7   (315)   3   (316)   7

(317)   4   (318)   2   (319)   6   (320)   3

(321)   2   (322)   5   (323)   1   (324)   4

(325)   2   (326)   3   (327)   4   (328)   5

(329)   5   (330)   6   (331)   6   (332)   5

(333)   2   (334)   7   (335)   9   (336)   4

(337)   3   (338)   9   (339)   2   (340)   5

(341)   4   (342)   5   (343)   4   (344)   4

345) 4	346) 6	347) 2	348) 2
349) 12	350) 7	351) 2	352) 10
353) 3	354) 4	355) 5	356) 1
357) 1	358) 6	359) 18	360) 11
361) 13	362) 1	363) 15	364) 5
365) 13	366) 25	367) 2	368) 12
369) 5	370) 1	371) 12	372) 5
373) 9	374) 12	375) 20	376) 10
377) 4	378) 23	379) 17	380) 6
381) 3	382) 10	383) 6	384) 3
385) 20	386) 8	387) 19	388) 11
389) 18	390) 5	391) 16	392) 11
393) 5	394) 6	395) 3	396) 9
397) 1	398) 10	399) 8	400) 8
401) 7	402) 3	403) 13	404) 3
405) 22	406) 20	407) 18	408) 19
409) 12	410) 3	411) 8	412) 17

www.math-knots.com

(413)	13	(414)	8	(415)	16	(416)	21
(417)	12	(418)	10	(419)	12	(420)	7
(421)	6	(422)	12	(423)	12	(424)	8
(425)	7	(426)	1	(427)	9	(428)	13
(429)	8	(430)	32	(431)	6	(432)	24
(433)	7	(434)	7	(435)	9	(436)	3
(437)	4	(438)	9	(439)	15	(440)	5
(441)	18	(442)	45	(443)	13	(444)	8
(445)	32	(446)	33	(447)	4	(448)	23
(449)	21	(450)	26	(451)	6	(452)	15
(453)	19	(454)	35	(455)	22	(456)	2
(457)	18	(458)	26	(459)	20	(460)	34
(461)	58	(462)	13	(463)	5	(464)	40
(465)	27	(466)	26	(467)	21	(468)	17
(469)	31	(470)	8	(471)	12	(472)	13
(473)	16	(474)	12	(475)	43	(476)	8
(477)	8	(478)	32	(479)	37	(480)	24

www.math-knots.com

(481)	17	(482)	3	(483)	39	(484)	46
(485)	18	(486)	2	(487)	15	(488)	10
(489)	20	(490)	8	(491)	49	(492)	37
(493)	1	(494)	27	(495)	15	(496)	26
(497)	15	(498)	11	(499)	13	(500)	10
(501)	19	(502)	24	(503)	1	(504)	50
(505)	1	(506)	18	(507)	16	(508)	12
(509)	1	(510)	44	(511)	48	(512)	34
(513)	19	(514)	23	(515)	40	(516)	18
(517)	22	(518)	34	(519)	14	(520)	30
(521)	38	(522)	7	(523)	30	(524)	41
(525)	27	(526)	144	(527)	198	(528)	224
(529)	208	(530)	272	(531)	108	(532)	48
(533)	112	(534)	390	(535)	36	(536)	105
(537)	105	(538)	60	(539)	96	(540)	360
(541)	54	(542)	80	(543)	304	(544)	120
(545)	380	(546)	90	(547)	24	(548)	684

(549) 30	(550) 117	(551) 608	(552) 60
(553) 180	(554) 60	(555) 160	(556) 156
(557) 36	(558) 200	(559) 90	(560) 140
(561) 140	(562) 234	(563) 84	(564) 60
(565) 108	(566) 120	(567) 72	(568) 36
(569) 175	(570) 60	(571) 165	(572) 120
(573) 24	(574) 120	(575) 105	(576) 36
(577) 312	(578) 72	(579) 264	(580) 1365
(581) 40	(582) 72	(583) 570	(584) 252
(585) 210	(586) 75	(587) 150	(588) 140
(589) 760	(590) 252	(591) 66	(592) 70
(593) 96	(594) 231	(595) 306	(596) 216
(597) 264	(598) 104	(599) 210	(600) 180
(601) 160	(602) 270	(603) 42	(604) 78
(605) 189	(606) 90	(607) 168	(608) 120
(609) 28	(610) 90	(611) 180	(612) 140
(613) 160	(614) 192	(615) 255	(616) 88

www.math-knots.com

(617) 150	(618) 150	(619) 325	(620) 1160
(621) 336	(622) 390	(623) 144	(624) 182
(625) 78	(626) 1683	(627) 240	(628) 156
(629) 4312	(630) 72	(631) 70	(632) 84
(633) 252	(634) 228	(635) 180	(636) 715
(637) 84	(638) 120	(639) 264	(640) 420
(641) 264	(642) 190	(643) 693	(644) 294
(645) 400	(646) 100	(647) 990	(648) 2244
(649) 132	(650) 200	(651) 192	(652) 280
(653) 216	(654) 240	(655) 114	(656) 1860
(657) 455	(658) 380	(659) 300	(660) 315
(661) 420	(662) 288	(663) 420	(664) 360
(665) 672	(666) 360	(667) 300	(668) 405
(669) 168	(670) 2296	(671) 450	(672) 240
(673) 270	(674) 336	(675) 126	(676) 180
(677) 546	(678) 48	(679) 320	(680) 638
(681) 400	(682) 480	(683) 224	(684) 1666

(685)	608	(686)	480	(687)	648	(688)	288
(689)	104	(690)	168	(691)	102	(692)	792
(693)	276	(694)	264	(695)	240	(696)	504
(697)	516	(698)	480	(699)	18165	(700)	2280
(701)	1386	(702)	5600	(703)	480	(704)	720
(705)	2835	(706)	2520	(707)	1344	(708)	3312
(709)	1050	(710)	20570	(711)	336	(712)	2340
(713)	1872070	(714)	1980	(715)	1470	(716)	676
(717)	480	(718)	11625	(719)	528	(720)	7800
(721)	3546375	(722)	630	(723)	3822	(724)	1188
(725)	35952	(726)	26703	(727)	11466	(728)	600
(729)	740	(730)	540	(731)	3564	(732)	66456
(733)	600	(734)	3000	(735)	360	(736)	1452
(737)	1176	(738)	1080	(739)	930	(740)	420
(741)	28000	(742)	1968	(743)	800	(744)	468
(745)	42920	(746)	1920	(747)	3740	(748)	505164
(749)	840	(750)	2184	(751)	22110	(752)	1134

(753) 2850	(754) 38896	(755) 17400	(756) 1020
(757) 384	(758) 564	(759) 1170	(760) 3520
(761) 1560	(762) 810	(763) 2184	(764) 900
(765) 870	(766) 4347	(767) 1800	(768) 456
(769) 1860	(770) 528	(771) 800	(772) 576
(773) 249642	(774) 2394	(775) 123984	(776) 216
(777) 588	(778) 980	(779) 924	(780) 207060
(781) 11^{13}	(782) 3^{11}	(783) 9^4	(784) 9^{11}
(785) 15^{13}	(786) 4^7	(787) 18^{13}	(788) 18^{11}
(789) 13^{11}	(790) 13^4	(791) 18^{18}	(792) 3^{15}
(793) 2^{10}	(794) 4^{13}	(795) 19^7	(796) 19^8
(797) 13^{15}	(798) 10^{11}	(799) 3^6	(800) 19^6
(801) 14^2	(802) 11^{11}	(803) 17^9	(804) 7^{16}
(805) 19^4	(806) 7^{10}	(807) 17^{12}	(808) 4
(809) 15^{10}	(810) 19^{14}	(811) 11^{13}	(812) 15^{10}
(813) 6^{10}	(814) 5^9	(815) 7^{18}	(816) 7^{16}
(817) 3^{17}	(818) 10^{10}	(819) 12^2	(820) 20^5

(821) 19^9 (822) 18^{14} (823) 13^{11} (824) 14^4

(825) 19^{12} (826) 17^{12} (827) 2^{12} (828) 8^{13}

(829) 19^{11} (830) 11^{10} (831) 20^{20} (832) 14^7

(833) 19^{14} (834) 2^7 (835) 8^8 (836) 2^{15}

(837) 8^2 (838) 6^3 (839) 6^{14} (840) 8^7

(841) 12^8 (842) 2^{14} (843) 12^{13} (844) 11^6

(845) 2^6 (846) 4^{12} (847) 10^{13} (848) 2^9

(849) 11^6 (850) 12^9 (851) 15^{16} (852) 5^7

(853) 13^{13} (854) 17^{14} (855) 5^{10} (856) 14^{15}

(857) 4^{17} (858) 19^{12} (859) 2^5 (860) 6^{10}

(861) 4^{10} (862) 11^{14} (863) 2^{18} (864) 9

(865) 4^6 (866) 3^{12} (867) 1   (868) 8^6

(869) 5^3 (870) 8^4 (871) 2^9 (872) 8^3

(873) 4^6 (874) 5^8 (875) 7^6 (876) 6^{12}

(877) 6^9 (878) 3^8 (879) 4^{12} (880) 6^6

(881) 7^8 (882) 7^3 (883) 2^8 (884) 3^9

(885) 5^6 (886) 3^4 (887) 8^4 (888) 5^4

www.math-knots.com

(889) 2^3

(890) 2^{12}

(891) 2^6

(892) 7^9

(893) 7^4

(894) 5^{12}

(895) 6^8

(896) 6^2

(897) 3^{12}

(898) 3^8

(899) 8^9

(900) 3^3

(901) 5^{12}

(902) 1

(903) 2^6

(904) 3^6

(905) 8^8

(906) 7^2

(907) 1

(908) 1

(909) 6^3

(910) 4^4

(911) 3^6

(912) 5^8

(913) 4^8

(914) 1

(915) 6^4

(916) 5^6

(917) 6^6

(918) 2^4

(919) 5^4

(920) 4^4

(921) 3^2

(922) 7^4

(923) 3^{16}

(924) 6^4

(925) 7^8

(926) 7^6

(927) 6^8

(928) 8^{12}

(929) 8^6

(930) 2^2

(931) 3^4

(932) 4^{12}

(933) 4^3

(934) 2^{16}

(935) 5^9

(936) $\dfrac{1}{2^{10}}$

(937) $\dfrac{1}{12^3}$

(938) $\dfrac{1}{3^4}$

(939) 10^2

(940) $\dfrac{1}{4}$

(941) 13

(942) 7^4

(943) 14

(944) 10^3

(945) $\dfrac{1}{17^3}$

(946) $\dfrac{1}{15}$

(947) 9^{10}

(948) $\dfrac{1}{7^6}$

www.math-knots.com

(949) 2^2

(950) 1

(951) 1

(952) $\dfrac{1}{2}$

(953) 10^3

(954) 19^2

(955) 18^3

(956) $\dfrac{1}{10}$

(957) $\dfrac{1}{16}$

(958) 10^6

(959) $\dfrac{1}{15^7}$

(960) 12

(961) $\dfrac{1}{20}$

(962) $\dfrac{1}{11^4}$

(963) $\dfrac{1}{17^2}$

(964) $\dfrac{1}{19^2}$

(965) 1

(966) $\dfrac{1}{14}$

(967) 20^9

(968) 20

(969) $\dfrac{1}{8^9}$

(970) $\dfrac{1}{12^7}$

(971) $\dfrac{1}{10^3}$

(972) 14^7

(973) 9^9

(974) $\dfrac{1}{19}$

(975) 10^5

(976) 4^2

(977) $\dfrac{1}{2^5}$

(978) 17^3

(979) $\dfrac{1}{2}$

(980) 11^5

(981) $\dfrac{1}{20}$

(982) $\dfrac{1}{14^4}$

(983) 10^3

(984) $\dfrac{1}{2^5}$

(985) 8^2 (986) 7^6 (987) 8^2 (988) 10^4

(989) $\dfrac{1}{14^8}$ (990) $\dfrac{1}{16}$ (991) 3^6 (992) 5^2

(993) $\dfrac{1}{6^4}$ (994) $\dfrac{1}{19^6}$ (995) $\dfrac{1}{10}$ (996) $\dfrac{1}{15^5}$

(997) $\dfrac{1}{12^6}$ (998) $\dfrac{1}{12}$ (999) $\dfrac{1}{19}$ (1000) $\dfrac{1}{4^2}$

(1001) 4 (1002) $\dfrac{1}{2}$ (1003) $\dfrac{1}{20^4}$ (1004) 4^9

(1005) 12^3 (1006) 1 (1007) $\dfrac{1}{20^2}$ (1008) $\dfrac{1}{5^3}$

(1009) $\dfrac{1}{19}$ (1010) 16 (1011) $\dfrac{1}{13^3}$ (1012) 5

(1013) 13^2 (1014) $\dfrac{1}{11^2}$ (1015) 8^8 (1016) $\dfrac{1}{5^{10}}$

(1017) $\dfrac{1}{10^5}$ (1018) 13^7 (1019) $\dfrac{1}{4^3}$ (1020) $\dfrac{1}{3^7}$

www.math-knots.com